THE PYTHAGOREAN THEOREM

for Babies

By Fred Carlson

The Pythagorean Theorem for Babies

First Edition

© 2013, Fred Carlson

ISBN-13: 978-1482000580

ISBN-10: 148200058X

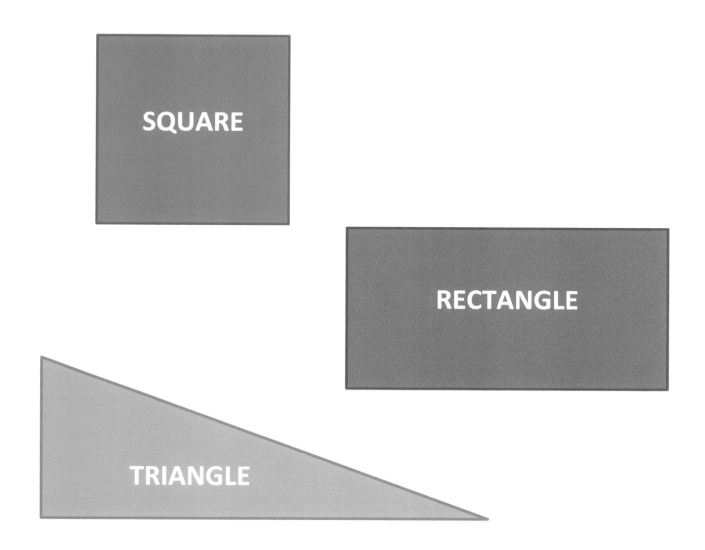

This book is about shapes like **SQUARES**, **RECTANGLES**, and **TRIANGLES**, and the **amazing** way they are all related.

This is a **SQUARE**.

A **SQUARE** has four sides and four corners that are all the same.

This block is a **SQUARE**.

SQUARES can be any size you want, as long as all of the sides and corners are the same.

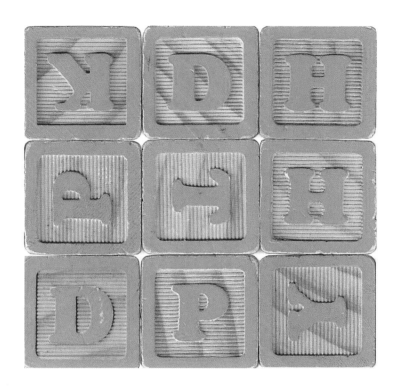

This **SQUARE** is made of nine blocks.

Each side is three blocks long.

This is a **RECTANGLE**.

A **RECTANGLE** has four corners that are all the same,

but the sides can be different.

This **RECTANGLE** is made of twelve blocks.

The sides are not all the same.

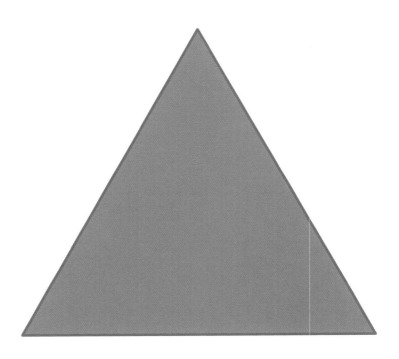

This is a **TRIANGLE**.

A **TRIANGLE** has three sides and three corners.

There are many ways to make a **TRIANGLE**,

but this book is about a special kind of **TRIANGLE**.

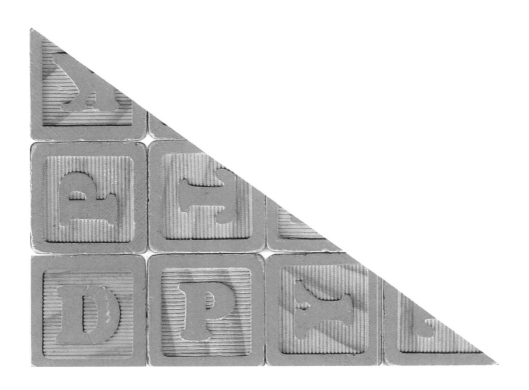

If you cut a **RECTANGLE** in half like this,

you get a shape called a **RIGHT TRIANGLE**.

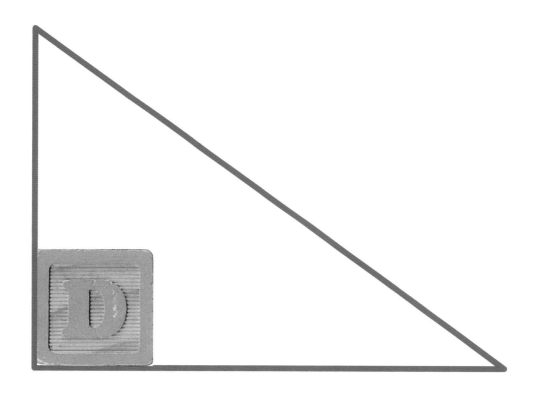

A **RIGHT TRIANGLE** has three sides, just like any other **TRIANGLE**,

but it's special because a **SQUARE** fits perfectly into one of the corners.

If you use blocks to make a **SQUARE** on each side of a **RIGHT TRIANGLE**, you just might notice something really neat.

(It's okay if you need help counting the blocks, smart baby.)

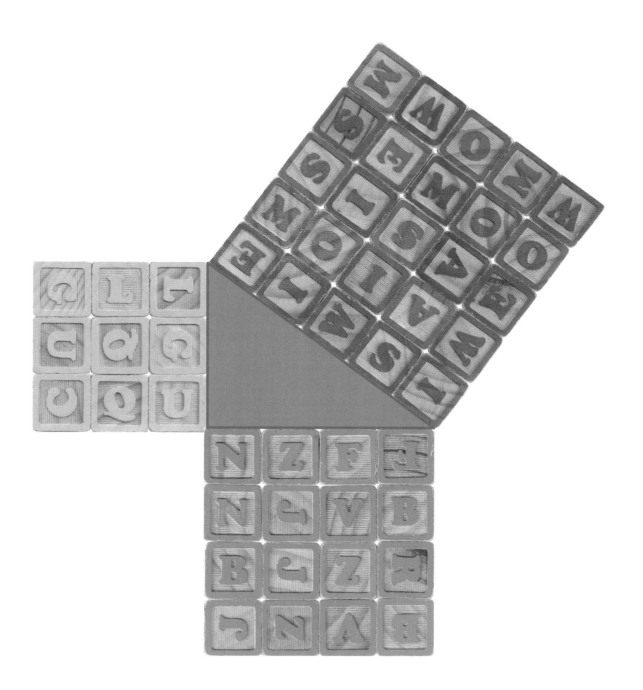

Nine Blocks   *plus*   *Sixteen Blocks*

The number of blocks in the two smaller **SQUARES**...

equals Twenty-Five Blocks!

... is the same as the number of blocks in the biggest **SQUARE**.

The Pythagorean Theorem says that is true for any **RIGHT TRIANGLE**.

A Theorem is something we can prove to be true.

Can you prove that the Yellow and Green **SQUARES**

are the same size as the Red **SQUARE**?

Four Blue **TRIANGLES** make a big Blue **SQUARE**.

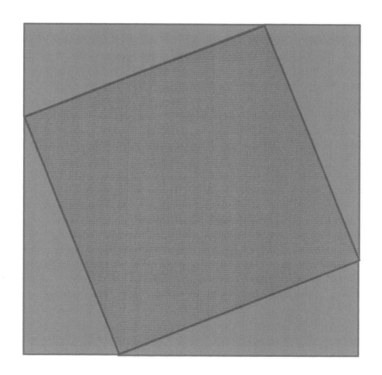

And the Red **SQUARE** fits perfectly inside!

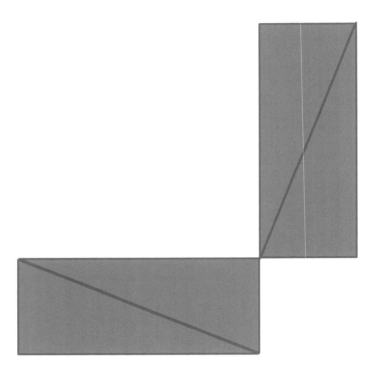

Four Blue **TRIANGLES** also make this funny shape.

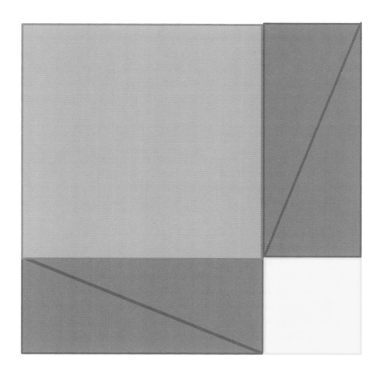

And the Yellow and Green **SQUARES** fit perfectly!

These shapes are both **SQUARES**, and they are both the same size.

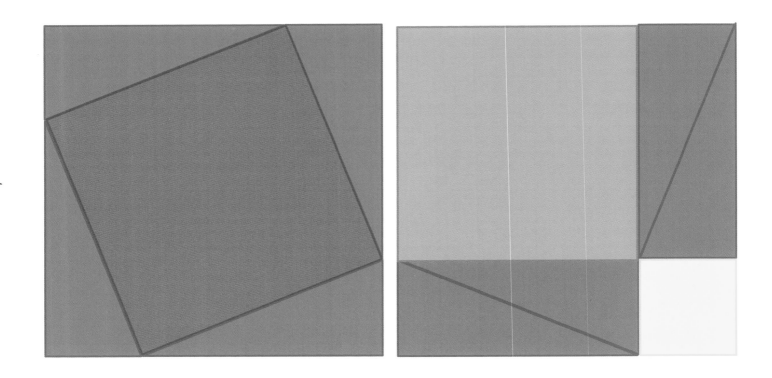

If you take away the Blue **TRIANGLES**, the shapes left over

must also be the same size.

That means that the Yellow and Green **SQUARES** are the

same size as the Red **SQUARE**, and that's just what we wanted to show!

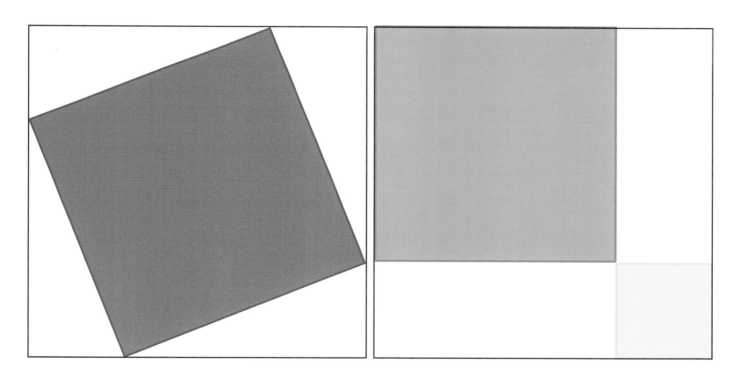

Congratulations, Baby, on your very first proof!

(I hope.)

Also in the Math for Babies Series:

Non-Euclidean Geometry for Babies

FRED CARLSON is a mathematician and an actuary. His understanding of The Pythagorean Theorem, and many other advanced mathematical concepts, is at least at the first grade level. If you have any questions or comments on this book or other forthcoming titles in the Math for Babies series, please contact Fred at mathforbabies@gmail.com.

34086124R00017

Made in the USA
Lexington, KY
22 July 2014